STEAM & Me™

ROBOTS

DINAH WILLIAMS

Starry Forest Books

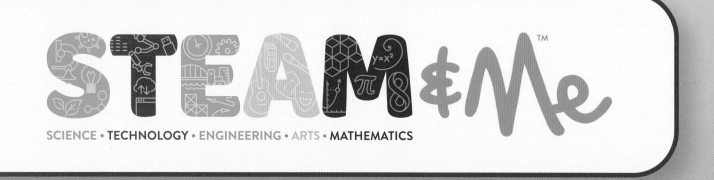

STEAM & Me

SCIENCE • TECHNOLOGY • ENGINEERING • ARTS • MATHEMATICS

Draw a super-smart robot. Create your own wind energy. Find out if your teeth are as sharp as a shark's. Go back in time to the world of dinosaurs or rocket into space. Power up that scientific brain of yours with **STEAM & Me**!

Photos, facts, and fun hands-on activities fill every book. Explore and expand your world with science, technology, engineering, arts, and math.

STEAM&Me is all about you!

Great photos to help you get the picture

New ideas sure to change how you see your world

Since the 1960s, car companies have used robots for jobs like welding and painting.

Give this robot a job!

Robots don't get tired or hungry or sick. They also don't get bored. This makes robots good at doing certain types of difficult work. Some robots work in factories, lifting heavy boxes or cutting pieces of metal. Robots can do the same job over and over without making mistakes.

Heavy Lifting

Kiva robots are only about 16 inches tall. But a Kiva weighs 320 pounds, as much as a huge gorilla. These robots are strong enough to lift packages up to 700 pounds. That's more than twice their own weight!

Get out a pencil and four pieces of paper. Choose a simple shape, like a circle or a square. Try to draw the exact same shape in the exact same spot on each piece of paper. Now look at them next to each other. Do they look identical? Most people can't do the same job over and over again and have it turn out perfectly, but some robots can!

Fascinating facts to fill and thrill your brain

Hands-on activities to spark your imagination

Beep-bop-boop!

Do you picture a robot as a boxy metal machine on legs with blinking lights? Robots come in many different shapes, sizes, and colors. And they can do many different tasks. Right this minute, robots are delivering packages, making sandwiches, lifting bales of hay, and vacuuming carpets. Robots often do jobs that people cannot do, like exploring the far reaches of space. In this book, you'll learn about what robots are, how they work, and what they can do.

Ready, set, robot! In boxes like these throughout the book, you'll do fun activities and experiments to help you understand more about robots and how they help us.

STEAM⋆Me

Robots can do everything from building huge rocket ships to picking fragile fruits.

Most robots don't look like people, but some do. The Kodomoroid robot looks human and reads the news on TV in Japan.

Machines Need Help

A toaster is a machine, but it isn't a robot. It can't do its job without a person to help it. Can you think of other machines in your house that are not robots?

What is a robot?

Robots are machines that can do things without human help. People design and build robots for all kinds of jobs, and sometimes just for fun. Some are so tiny they can swim in your **bloodstream** to fight disease. Others are huge. There is even a 51-foot-long dragon robot that breathes fire!

Brains of Steel

A small computer in robots tells them what to do and how to do it. It works kind of like your brain does.

STEAM&Me

To control robots, scientists use computer **programs**. A program tells robots what to do by giving them steps to follow. Write a program to make your favorite sandwich. Describe each step in order. You can ask a grown-up to help you write it down. Now ask someone to follow your program and see if they make the sandwich just right!

Move it, robot!

All robots have moving parts. Some robots stay in one place and do one job over and over again, like drilling perfect holes in sheets of metal. Others can move from one place to another. Depending on its job, a robot may need wheels, tracks, or legs to get around.

Robots by Nature

Some robots move like animals. Others have many legs and move like bugs. What kind of creature does this robot look like?

Robots that can move like people take humans' places doing dangerous work. They can do fun things, too. The Atlas robot can run up stairs, jump over logs, and do a perfect backflip.

Super Speed

Some robots are made to move quickly. One of the fastest running robots is the WildCat robot, which gallops like a horse and can run outside at nearly 20 miles per hour.

Some robots work closely with humans. These robots use lasers and cameras so they'll know when a person is nearby and will stop working to keep their human coworker safe.

Have you ever tried to walk around with your eyes closed? With a grown-up nearby, find a place where you can safely move around. Close your eyes and try to find your grown-up. What senses do you use to locate the grown-up?

STEAM&Me

Robots make sense with sensors.

When you cross a street, you use your senses of sight and hearing to look and listen. This is how you know if cars are coming. Robots use their **sensors**, which are like your eyes and ears. Sensors help robots create a picture of their world. Robots use different sensors for different jobs. Camera and **laser** sensors alert robots to movement. Microphones are sensors that can help robots process sounds.

Lights Out

Some robot sensors act like senses found in animals. The wheeled Robat moves in the dark. Like the bat pictured here, it finds its way by bouncing sounds off objects. These sounds can't be heard by human ears.

Robots to the Rescue!

Scientists are building snake-bots that can slither through tight spots to help people who are stuck under rocks or collapsed buildings after avalanches or earthquakes.

NASA is the US government agency that explores space. NASA has built robots that can go deep inside fiery, active volcanoes. These robots help scientists learn how volcanoes explode. One day, these robots may explore volcanoes on other planets, too.

Robots can take big risks.

Have you ever climbed down into an active volcano or swum under the ice? Better not! Those places are too dangerous for humans. They are too hot, too cold, or too deep underwater. Robots can be designed to go where people can't go.

It's a bird! It's a plane! It's a robot!

People can't fly on their own. But some robots can. Some small flying robots have wings that look like they belong on a bug. Larger robots fly using propellers, like the ones found on some **drones**. Sometimes people steer a flying robot with a remote control. Other times the robot flies on its own, using a program in its computer brain.

Drones fly into natural disasters like wildfires, floods, or tornadoes. They send pictures that help locate trapped people. They help rescuers figure out where to go.

Buzz, Buzz!

Some robots are tiny. One of the world's smallest robots is the RoboBee, which can fly in the air and go underwater to one day help with search and rescue missions.

Some robots can work underwater.

More than 70 percent of Earth is covered by water. Robots don't need to breathe, so they can stay underwater as long as needed. They can dive to the deepest parts of the ocean, where it is too cold and dangerous for people. The information that underwater robots collect helps scientists do things like create maps of the ocean floor and discover new deep-sea creatures.

It's Cold Down Here!

The funny-named robot Boaty McBoatface swam 60 miles under an ice sheet in Antarctica to gather information about the water there for scientists.

ROPOS are remote-controlled robots that can dive more than 3 miles deep into the ocean to take photos and collect samples.

STEAM

Fill a big bowl or bucket with water. Gather a few toys and objects that can go in the water. Try to find something made of plastic, something made of metal (like a spoon), and something made of wood. Place them in the water. Do they sink or float? Which material would be best for a robot that needs to go under the water? Which material would be best for a robot that needs to travel on top of the water?

Meet Pepper—
Pepper is a smart
robot that can learn
about you, talk to
you, and understand
your feelings.

Happy and You Know It

Sophia, an artificial intelligence robot, has a face that looks like a real person. It can even show human feelings.

Some robots are trained to think like you.

Quick, what's 423,987 + 245,987? That's easy for some robots, but even the fastest robots need humans to teach them. Scientists write programs that help robots get smarter. When a robot can learn lessons and get better at doing jobs, it's working like a human brain in some ways. We say that kind of robot has **artificial intelligence**, or A.I.

I'm Hungry

YuMi can do lots of jobs. It can even cook for you! Another robot makes lunch every day at CaliBurger restaurants.

Heavy Lifting

Kiva robots are only about 16 inches tall. But a Kiva weighs 320 pounds, as much as a huge gorilla. These robots are strong enough to lift packages up to 700 pounds. That's more than twice their own weight!

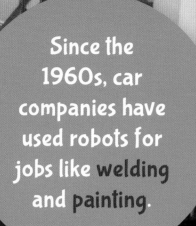

Since the 1960s, car companies have used robots for jobs like **welding** and **painting**.

Give this robot a job!

Robots don't get tired or hungry or sick. They also don't get bored. This makes robots good at doing certain types of difficult work. Some robots work in factories, lifting heavy boxes or cutting pieces of metal. Robots can do the same job over and over without making mistakes.

STEAM

Get out a pencil and four pieces of paper. Choose a simple shape, like a circle or a square. Try to draw the exact same shape in the exact same spot on each piece of paper. Now look at them next to each other. Do they look identical? Most people can't do the same job over and over again and have it turn out perfectly, but some robots can!

Robots can have fun, too.

Not all robots are made to do work. Some robots are made just to play, talk, and dance. You can find them in museums, restaurants, and amusement parks.

Sporty Robots

People build robots that can play sports like soccer, tennis, baseball, and ping-pong. Some are so good they can beat human players!

Sit, Robot

Some toy robots look and act like real pets and can learn tricks.

Think about the kind of robot you might like to build. Would it dance? Sing? Meow like a kitten? Make you a snack? Draw a cartoon or comic strip that shows what your robot would do.

Disneyland has used robots since 1965. The robots look like birds, presidents, pirates, and movie characters that have come to life.

Tiny robots called nanobots are being built to fight disease from inside a person's body.

Calling Dr. Robot!

What if your doctor was a robot? Robots are already helping doctors in lots of ways. Some help surgeons operate inside a patient's body. Other robots give out medicine and remind people of their doctor's directions.

Open Wide!

People studying to be doctors and dentists sometimes practice on robots.

Robot Skeleton

People who have been badly hurt and can't walk can use robotic suits to help them walk again.

Robots Inside and Out

Robots can help inside the house, but also do lots of work outside. Robots can play fetch with your dog and even mow the lawn for you.

Welcome home, robot.

Robots do lots of hard and dangerous jobs. But they can also do your chores! The Aeolus robot can deliver food, pick up around the house, and find things you lost. It's a smart robot: the more it learns, the more helpful it becomes.

Floor-cleaning robots use **sensors** to find their way around rooms.

Someone's in the Kitchen

Some robots can cook entire meals and even clean up.

Robots make great astronauts.

Traveling to Mars would take you seven months. When you got there, you would be chilly. The coldest temperatures on Mars are even colder than Antarctica. There's also no air on Mars. You might not be able to live there, but a robot can! The space program NASA has sent robots called **rovers** to Mars to explore the planet.

Earth to Robot!

Floating robots like the JEM Internal Ball Camera gather images and videos for the crew on the International Space Station.

NASA's Mars Curiosity rover has long robotic arms. How do robotic arms work? Like yours! Slowly pick up a small toy and watch carefully how your fingers, arm, and hand move to grab it. Did your fingers open and close? Did your elbow bend? Did your arm move up and down? Scientists need to think about all these tiny movements to design robotic arms.

STEAM & Me

Curiosity test rover

Spirit and Opportunity test rover

Sojourner flight spare

Spacecraft engineers test rovers on Earth before they go to Mars. The Sojourner rover was the size of a microwave. The Curiosity rover landed on Mars in 2012, and it is 10 feet long!

Robots can do so many things.

People use robots more every year. Scientists and engineers are always inventing new ones. Which of these robots does what you want a robot to do? Which one do you think is the most helpful kind of robot? What other sorts of things could a robot help you do? Invent the coolest new robot you can imagine. Draw a poster featuring your robot and listing all the things it can do.

Glossary

Learn these key words and make them your own!

artificial intelligence: the power of a machine to think and learn like a person. *A robot that can learn from its mistakes has* artificial intelligence.

bloodstream: the blood that moves through the body. *Your* bloodstream *moves blood from your heart all through your body and back again.*

drone: an aircraft that flies without a person on board. *The small* drone *flew high into the air to take photos of the river below.*

laser: a device that makes a strong beam of light. *When a robot's* laser *hits an object, the robot can tell that the object is there.*

program: a set of steps written by people, telling a computer what to do. *A computer scientist writes a* program *that teaches a computer to play checkers.*

rover: a space-exploring robot that works away from Earth. *NASA has sent several* rovers *to Mars.*

sensor: a device that measures changes in heat, light, movement, or sound. *Some robots use motion* sensors *to know when someone is walking near them.*

STEAM & Me and Starry Forest® are trademarks or registered trademarks of Starry Forest Books, Inc. • Text and Illustrations © 2020 and 2021 by Starry Forest Books, Inc. • This 2021 edition published by Starry Forest Books, Inc. • P.O. Box 1797, 217 East 70th Street, New York, NY 10021 •
ISBN 978-1-946260-89-5 • Manufactured in China • Lot #: 2 4 6 8 10 9 7 5 3 1 • 03/21

ASP: Alamy Stock Photo; GI: Getty Images; SS: Shutterstock. Cover, PaO_STUDIO/SS; 5, Suwin/SS; 6, David L. Moore - JPN/ASP; 6, (LO) New Africa/SS; 7, BigBlueStudio/SS; 8, PA Images/ASP; 8-9, John Tlumacki/The Boston Globe via GI; 9, Boston Dynamics; 10, Suwin/SS; 11, Rudmer Zwerver/SS; 12, NASA; 12-13, NASA/JPL-Caltech; 14-15, Bureau of Land Management; 15, Wyss Institute at Harvard University; 16, National Oceanography Centre; 17, S.Bachstroem/SS; 18, MikeDotta/SS; 18, (LO) Anton Gvozdikov/SS; 19, PaO_STUDIO/SS; 20, David Paul Morris/Bloomberg via GI; 20-21, TRAIMAK/SS; 22, (UP) Tinxi/SS; 22, (LO) quangmooo/SS; 23, Whitenep/Wikimedia Commons; 24, Volodymyr Horbovyy/SS; 25, (UP) Zapp2Photo/SS; 25, (LO) Ivan Chudakov/SS; 26, Paolo Bona/SS; 26-27, Daniel Krason/SS; 27, dpa picture alliance/ASP; 28, JAXA/NASA; 29, NASA/JPL-Caltech; 30, (UP) David Paul Morris/Bloomberg via GI; 30, (CTR) Paolo Bona/SS; 30, (LO) John Tlumacki/The Boston Globe via GI; 31, (UP LE) Bureau of Land Management; 31, (CTR LE) PA Images/ASP; 31, (LO LE) Wyss Institute at Harvard University; 31, (UP RT) Suwin/SS; 31, (CTR RT) PaO_STUDIO/SS; 31, (LO RT) Boston Dynamics; 32, (CTR) Jenson/SS; 32, (LO) VTT Studio/SS; Back cover, (UP) quangmooo/Shutterstock, (LO LE) PA Images/ASP, (LO CTR) Castenoid/Adobe Stock